U.S. Department
of Transportation

National Highway
Traffic Safety
Administration

People Saving People
http://www.nhtsa.dot.gov

DOT HS 809 724
NHTSA Technical Report

March 2004

Evaluation of Rear Window Defrosting and Defogging Systems

This document is available to the public from the National Technical Information Service, Springfield, Virginia 22161.

1. Report No. DOT HS 809 724	2. Government Accession No.	3. Recipient's Catalog No.	
4. Title and Subtitle Evaluation of Rear Window Defrosting and Defogging Systems		5. Report Date March 2004	
		6. Performing Organization Code	
7. Author(s) Christina Morgan		8. Performing Organization Report No.	
9. Performing Organization Name and Address Evaluation Division; Office of Planning, Evaluation and Budget National Highway Traffic Safety Administration Washington, DC 20590		10. Work Unit No. (TRAIS)	
		11. Contract or Grant No.	
12. Sponsoring Agency Name and Address Department of Transportation National Highway Traffic Safety Administration Washington, DC 20590		13. Type of Report and Period Covered NHTSA Technical Report	
		14. Sponsoring Agency Code	
15. Supplementary Notes			

16. Abstract

Rear window defrosting and defogging systems are not required on motor vehicles by any Federal standard. Rear window defoggers became available as optional or standard equipment in most cars during the 1970's or 1980's and are popular with consumers. Today, almost all passenger cars, minivans, and sport utility vehicles have rear window defoggers, but most pickup trucks and full-size vans do not.

The analysis examined whether there were proportionately fewer backing-up and changing-lane crashes involving cars with rear-window defoggers than cars without rear-window defoggers. The database was extracted from State crash files. The analyses did not show a benefit for rear window defoggers. The main analysis found that rear window defoggers have no effect on changing lane and backing crashes in conditions when they are most likely used (when raining or snowing, during the earlier part of the morning, or during winter).

Even though the statistical analyses did not show a significant reduction in actual crashes, we would expect most drivers to like rear-window defoggers because they are quite convenient, and improved rearward vision helps a driver feel more mobile and secure when changing lanes or backing up.

17. Key Words rear window defrosting and defogging systems; statistical analysis; evaluation; State crash files		18. Distribution Statement		
19. Security Classif. (Of this report) Unclassified	20. Security Classif. (Of this page) Unclassified		21. No. of Pages 49	22. Price

Form DOT F 1700.7 (8-72)   Reproduction of completed page authorized

TABLE OF CONTENTS

EXECUTIVE SUMMARY

Rear window defrosting and defogging systems, hereafter referred to as rear window defoggers, allow the driver to see through the rear window under adverse weather conditions. They get rid of condensation, frost, ice, and/or snow on the back windows. They are most likely to be used if any one or more of the following occur: it is snowing or raining, during the earlier part of the morning or during winter. A clear window will obviously help a driver who is backing up, and can also help a driver see if it is safe to change lanes.

Rear window defoggers are not required on motor vehicles by any Federal standard. Rear window defoggers became available as optional or standard equipment in most cars during the 1970's or 1980's and are popular with consumers. Today, almost all passenger cars, minivans, and sport utility vehicles have rear window defoggers, but most pickup trucks and full-size vans do not.

The analysis examined whether there were proportionately fewer backing-up and changing-lane crashes involving cars with rear-window defoggers than cars without rear-window defoggers. The basic analytical method was to estimate the overall reduction of rear-impact involvements to cars that had been backing or changing lanes just before the crash relative to a control group of rear-impact involvements to cars that were stopped prior to the crash in adverse conditions when defoggers are more likely to be used.

The analysis databases were initially extracted from State crash files of Florida, Maryland, Michigan and Pennsylvania. Ward's Automotive Yearbooks supplied information on the proportion of vehicles equipped with defoggers, by make-model and model year. However, we had no information whether specific, individual crash-involved vehicles were equipped with defoggers. The initial databases included 321,118 rear-impact cases from Florida (1986-1999), 74,725 cases from Maryland (1986-2000), 321,826 cases from Michigan (1981-1991) and 182,460 cases from Pennsylvania (1981-2000). But the analysis was limited to the Florida and Michigan data, because the Pennsylvania and Maryland data did not have an adequate number of changing lane and backing-up cases.

Number Of Cases in Initial Databases By Crash Type And State

State	Changing Lane & Backing Crashes	Stopped Crashes	Total
Michigan	71,668	250,158	321,826
Florida	44,761	276,357	321,118
Pennsylvania	7,791	174,669	182,460
Maryland	13,121	61,604	74,725

It was hard to detect a reduction in backing and changing lanes crashes, since rear window defoggers were gradually introduced into the cars. For example, there was not a single high-sales make-model that went from zero defoggers in one model year to 100 percent in the next, or even had close to such a jump. Logistic regression analyses were used to calibrate the ratio of backing and changing-lane crashes to control-group crashes as a function of the percentage of vehicles with defoggers, and also to control for other factors related to the driver, vehicle, crash and environmental circumstances that could affect the mix of relevant to non-relevant crashes and/or increase or decrease the effect of defoggers.

We were unable to conclude that rear window defoggers reduce crashes. The analyses did not show a benefit for rear window defoggers. The main analysis found that rear window defoggers have no effect on changing lane and backing crashes in conditions when they are most likely used (when raining or snowing, during the earlier part of the morning, or during winter).

Even though these data do not show a statistically significant crash reduction, we would still expect most drivers to like rear window defoggers because they improve rearward vision. They can reduce or eliminate the need to scrape the rear window from outside the vehicle after a frost, a snowfall, or freezing rain, and these systems can eliminate the need for periodic stopping while driving to scrape accumulating snow or wipe condensation from the rear window.

It should be noted that there are a large unknown number of backing up crashes of minor or no damage that are not reported to the police. For example, a driver backs out of the driveway and hits a pole, mailbox, or a neighbor's vehicle. It is possible that rear window defoggers are effective in reducing these unreported crashes. However, we have no data to evaluate these unreported crashes.

BACKGROUND

FMVSS 103 requires all passenger cars, multipurpose passenger vehicles, trucks and buses manufactured on or after January 1, 1968, to have windshield defrosting and defogging systems. However, rear window defrosting and defogging systems, hereafter referred to as rear window defoggers, are not required. Rear window defoggers became available as optional or standard equipment in most cars during the 1970's or 1980's and are popular with consumers. Rear window defoggers allow the driver to see through the rear window under adverse weather conditions. Clear vision is especially important when the driver wants to back up or change lanes.

FMVSS 104 requires all passenger cars, multipurpose passenger vehicles, trucks and buses manufactured on or after January 1, 1968, to have windshield wiping and washing systems. Similar to rear window defoggers, rear-window wiping and washing systems are not required. Rear window wipers have been installed on a smaller group of vehicles for even better vision.

This study assesses the effectiveness of rear window defoggers in reducing crashes in which a driver is backing up, changing lanes, or performing other tasks that would be facilitated by vision through the rear window. Typically, studies to assess the effectiveness of a vehicle safety standard consist of an analysis of crash data that compare the experience of vehicles that meet the safety standard to those that do not meet the standard. For most evaluation studies, we can identify exactly what vehicles were equipped with a safety technology, and which ones were not, based on the model year (e.g., before/after a standard's effective date, or before/after a date when the technology became standard equipment) or dedicated VIN characters (e.g., air bags), or specific mention in the crash data (e.g., child safety seats). However, rear-window defoggers and wipers have never been required safety equipment and were offered as optional equipment, for many years, on quite a few make-models. The presence of rear-window defoggers and wipers generally cannot be deduced from the first 12 characters of the VIN, and it is rarely (if ever) mentioned on crash data files. Therefore, it is necessary to obtain information on the proportion of vehicles with rear-window defoggers or wipers from another source.

REAR WINDOW DEFOGGER AND WIPER INFORMATION

Rear window defogger and wiper information is published yearly in the Ward's Automotive Yearbook[1]. Ward's Yearbook lists the percentage of factory-installed equipment on passenger vehicles by manufacturer and model. Each yearbook contains a list of factory-installed equipment for the previous years' production of vehicles. For example, Ward's Automotive Yearbook 2002 contains a list of factory-installed equipment for vehicles produced in calendar year 2001. Ward's groups the information about factory-installed equipment into four different vehicle groups: Domestic Cars, Imported Cars, Domestic Light Trucks, and Imported Light Trucks. Light trucks include all pickup trucks, vans, and Sport Utility Vehicles (SUVs). Table 1 shows the availability of rear window defogger and rear window wiper information contained in Ward's Automotive Yearbooks by the four vehicle groups.

---

[1] This book is published by Ward's Communications, Inc.

Table 1
Availability of Rear Window Defoggers and Wipers by Vehicle Group

Vehicle Group	Rear Defoggers	Rear Wipers
Domestic Cars	1973-later	1983-later
Imported Cars	1975-1998	1983-later
Domestic Light Trucks	1987-later	1987-later
Imported Light Trucks	1976-later	1983-later

Domestic cars include all cars produced in the U.S., Canada and Mexico for the U.S. market. They include some models sold by foreign-based manufacturers but assembled in North America. For example, the Honda Accord and Civic have been assembled in the U.S. since 1987 and are listed with domestic cars since 1987. Similarly, the domestic light trucks (pickups, vans and SUVs) include all light trucks produced in North America for the U.S. market, even some models sold by foreign-based manufacturers. For example, the Toyota Tundra, Nissan Quest, and the Honda Odyssey are a few of the models sold by foreign-based manufacturers included in the domestic light truck list.

Imported cars and imported light trucks include all vehicles imported for U.S. sales. Consequently, this can include some models by domestic-based manufacturers assembled abroad. For example, the Geo Storm, Ford Aspire, and Cadillac Catera are included in the imported car list.

Ward's classifies factory-installed equipment by calendar year of production. This is not exactly the same as model year, since some new model year vehicles are usually produced and introduced in the fall of the previous year. For example, vehicles produced in calendar year 2001 will include mostly 2001 model year vehicles and a small portion of 2002 model year vehicles. In this report, the calendar year production numbers are being used to represent the model year numbers (i.e., calendar year 2001 numbers are being used to represent model year 2001 model year numbers). This adds a little error to the analysis, but it is the only information available on rear window defoggers.

Ward's Automotive Yearbook reports the combined percentage of rear window defoggers that blow air on the rear window and that electrically heat the rear window for 1976 model year domestic cars. For 1977-1981 model year domestic cars, Ward's separately reports the percentage of rear window defoggers that blow air on the rear window and that electrically heat the rear window. It appears that rear window defoggers that blew air on the rear window were discontinued after 1981. The only 1981 model year domestic car with rear window defoggers that blew air was the Chevrolet Caprice. Only 1.6 percent of Chevrolet Caprice had rear window defoggers that blew air, but 44.3 percent of them had rear window defoggers that electrically heated the rear window. Since the percentage reported was a combination of both types of rear window defogger in 1976, the percentage used in the analysis for model year

1977-1981 will be the combination of both types. For 1982 model year domestic cars, Wards also reports the percentage of factory installed rear window washers.

Today, almost all new cars and imported light trucks sold in the U.S. have rear window defoggers, but only about half of the domestic light trucks have rear window defoggers. Table 2 shows the percentage of factory-installed rear window defoggers by model year and vehicle group.

Table 2
Percentage Factory-Installed Rear Defogger By Model Year And Vehicle Group

Model Year	Domestic Cars	Imported Cars	Domestic Light Trucks	Imported Light Trucks
1973	16.4%			
1974	21.5%			
1975	25.7%	72.1%		
1976	29.9%	67.4%		0.6%
1977	35.0%	82.4%		0.6%
1978	39.1%	88.0%		0.0%
1979	47.5%	87.6%		0.0%
1980	50.5%	86.1%		0.1%
1981	50.9%	97.6%		0.0%
1982	55.9%	99.5%		0.0%
1983	59.0%	97.2%		6.1%
1984	63.1%	97.5%		11.9%
1985	74.0%	95.2%		11.6%
1986	71.7%	91.8%		13.4%
1987	73.6%	89.9%	17.3%	18.6%
1988	81.7%	87.2%	19.2%	19.2%
1989	83.3%	90.4%	17.0%	29.1%
1990	81.2%	91.2%	25.6%	31.9%
1991	88.6%	87.8%	31.5%	47.8%
1992	90.8%	96.0%	32.5%	44.9%
1993	91.9%	92.0%	37.3%	56.5%
1994	93.3%	93.1%	35.5%	65.0%
1995	94.1%	94.4%	40.2%	71.2%
1996	94.3%	95.7%	47.2%	88.4%
1997	94.7%	96.2%	40.6%	93.7%
1998	95.4%	99.9%	50.8%	98.4%
1999	95.2%		48.8%	100.0%
2000	96.9%		50.0%	99.6%
2001	94.2%		54.0%	99.7%

The percentage of new vehicles sold in the U.S. with factory-installed rear window defoggers has steadily increased since Ward's started reporting rear window defoggers. For domestic cars, the rate has gone from 16 percent for model year 1973 cars to 94 percent of model 2001 cars. For imported cars, the rates have steadily increased like those of domestic cars, but are generally higher than those of domestic cars until recently, when almost all new cars sold in the United States are equipped with rear window defoggers. The rates for light trucks have some of the same general trends as cars. The rates have steadily increased like the car rates and the rates for imported light trucks are higher than those of domestic light trucks. The main difference between the car and light truck rates is that the domestic light truck rates do not catch up to the imported light truck rates in recent model years. Almost all imported light trucks have rear window defoggers, but only 54 percent of domestic light trucks have rear window defoggers.

Table 3 shows the percentage of factory-installed rear window defoggers by model year for six make-models of cars with high sales volumes.

In general, the percentage of factory-installed rear window defoggers on models by domestic-based manufacturers have steadily increased by model year, but the percentage on models by foreign-based manufacturers was high in the late 1970's and has remained high. Most of the time the percentage of rear window defoggers increases slightly from one year to the next for each make-model.

The percentage has never increased from 0 to 100 percent in consecutive model years for any make-model, but in several make-models the percentage has had a big positive jump in consecutive model years. The Ford Tempo, Ford Crown Victoria, and Honda Civic shown in Table 3 have had large jumps in the percentage of rear defoggers in consecutive model years.

Table 3
Percentage Of Factory-Installed Rear Window Defoggers
By Model Year For Six High-Sales Make-Models

Model Year	Chevrolet Cavalier	Ford Tempo	Ford Crown Victoria	Honda Civic	Toyota Corolla Domestic	Toyota Corolla Imported
1975				11.7%		95.0%
1976				12.9%		85.5%
1977				20.0%		85.0%
1978				91.0%		84.5%
1979			37.6%	87.1%		67.4%
1980			39.1%	70.6%		100%
1981			61.2%	79.6%		100%
1982	90.4%		64.5%	100%		100%
1983	51.1%	63.3%	75.3%	90.0%		100%
1984	49.8%	50.6%	63.3%	95.0%		100%
1985	47.9%	53.4%	80.1%	89%		100%
1986	55.2%	60.4%	80.9%	84.0%		100%
1987	62.0%	64.2%	86.7%	100%*		0%
1988	66.5%	68.6%	89.3%	100%*		100%
1989	61.1%	94.9%	87.5%	100%*	100%	100%
1990	61.0%	97.1%	89.4%	100%*	94.0%	100%
1991	70.1%	92.3%	94.0%	100%*	100%	100%
1992	66.8%	90.3%	95.3%	100%*	100%	100%
1993	73.1%	88.5%	97.1%	100%*	100%	100%
1994	75.2%	100%	97.7%	100%*	62.0%	62.3%
1995	77.4%		100%	100%*	68.0%	62.5%
1996	75.5%		100%	100%*	74.0%	78.0%
1997	80.1%		100%	100%*	75.0%	98.0%
1998	78.9%		100%	100%*	74.7%	
1999	85.3%		100%	100%*	79.0%	
2000	91.8%		100%	100%*	87.0%	
2001	100%		100%	100%*	46.8%	

* Percentage for both domestic and imported Civics

Table 4 shows the biggest positive jumps for several high-sales make-models, ordered by amount of gain. The Honda Civic had one of the largest positive jumps by any make-model. The percentage for rear window defoggers jumped 71 percentage points in the Civic from 20.0 percent in 1977 to 91.0 percent in 1978. About 30 make-models had an increase of 20 percentage points or more in one year. In the Tempo, the percentage of rear window defoggers jumped from 68.6 percent in 1988 to 94.9 percent in 1989. In the Grand Marquis, it jumped from 58.0 percent in 1980 to 83.6 percent in 1981. In the Crown Victoria, it jumped from 39.1 percent in 1980 to 61.2 percent in 1981. About 100 make-models had an increase of 10 percentage points or more in consecutive years. Many of the high-sales foreign make-models had rear window defoggers as standard equipment when these models were first introduced. The Honda Accord, Nissan Sentra, and Toyota Camry were introduced with rear window defoggers as standard equipment.

Table 4
Biggest Positive Jumps For Several High-Sales Make-Models

Make-Model	Model year	Percentage	Model Year	Percentage	Gain
Honda Civic	1977	20.0%	1978	91.0%	71.0%
Ford Tempo	1988	68.6%	1989	94.9%	26.3%
Mercury Grand Marquis	1980	58.0%	1981	83.6%	25.6%
Ford Crown Victoria	1980	39.1%	1981	61.2%	22.1%
Ford Escort	1985	42.2%	1986	60.0%	17.8%
Chevy Celebrity	1986	64.5%	1987	79.5%	15.0%
Chevy Chevette	1977	14.0%	1978	28.4%	14.4%
Pontiac Grand Am	1993	78.0%	1994	89.3%	11.3%
Chevy Cavalier	1990	61.0%	1991	70.1%	9.1%
Ford Taurus	1990	91.5%	1991	100%	8.5%
Olds Ciera	1990	74.5%	1991	82.8%	8.3%
Toyota Corolla	1999	79.0%	2000	87.0%	8.0%
Chevy Caprice	1985	63.5%	1986	70.5%	7.0%

Table 5 shows the percentage of factory-installed rear window defoggers by model year for eight popular make-models of light trucks. In general, the percentage of minivans and sport utility vehicles with rear window defoggers is much larger than pickup trucks. Recall in Table 2 that the rates for imported light trucks were generally higher than those of domestic trucks, but domestic light trucks did not catch up to the imports (Table 2). In other words, the difference in percentage of domestic and imported light trucks with rear defoggers in recent model years is larger than earlier model years. The difference appears to be caused by:

- Almost all recent model years of minivans and SUVs both domestic and imported have rear window defoggers,
- Very few recent model years of pickup trucks and full-sized vans have rear window defoggers, and
- A shift to assemble all pickup trucks in North America, even pickup trucks by foreign-based manufacturers.

This causes the overall percentage of rear window defoggers for imported light trucks to be much larger than that of domestic light trucks. In model year 2001, all imported light trucks had rear window defoggers but only half of the domestic light trucks had them.

Table 5
Percentage Of Factory-Installed Rear Window Defoggers
By Model Year For Eight Popular Light Truck Make-Models

Model Year	Pickup Trucks			Minivans		Sport Utility Vehicles		
	Chevrolet C/K 1500 Silverado	Ford F150 Sierra	Toyota Pickup - Tacoma	Dodge Caravan	Chevrolet Lumina - Venture	Ford Explorer	Chevrolet Blazer "S"	Toyota 4Runner
1987	3.4%	0%	0%	67.4%			61.5%	66.8%
1988	2.6%	0%	0%	75.9%			55.9%	82.4%
1989	1.5%	0%	0%	79.5%			0%	100%
1990	1.3%	0%	0%	86.1%			51.4%	0%
1991	1.4%	0%	0%	93.1%	72.5%	83.7%	78.5%	97.0%
1992	0.8%	0%	0%*	89.1%	78.9%	96.7%	86.4%	96.0%
1993	1.0%	0%	0%*	86.9%	74.5%	98.3%	93.9%	96.9%
1994	1.6%	0%	0%*	87.3%	92.2%	99.0%	96.8%	99.0%
1995	2.5%	0%	0%*	85.8%	94.0%	99.3%	100%	99.0%
1996	2.5%	0%	0%	92.5%	92.6%	99.3%	100%	97.0%
1997	2.7%	0%	0%	92.1%	97.6%	99.4%	100%	98.0%
1998	2.9%	0%	0%	93.8%	97.8%	92.8%	97.1%	100%
1999	3.0%	0%	0%	89.0%	100%	100%	98.0%	100%
2000	12.6%	0%	0%	92.1%	100%	100%	96.7%	100%
2001	18.4%	13.6%	0%	93.4%	100%	100%	97.2%	100%

* Percentage for both domestic and imported

Table 6 shows the percentage of factory-installed rear window wipers by model year and vehicle group. The percentages of factory-installed rear window wipers for domestic cars are fairly consistent for the last 20 model years. The percentages for imported cars are also fairly consistent but only for the last 10 model years. The rates for imported cars are also slightly larger than those of domestic cars, but neither of the rates for domestic or imported cars is very large.

By contrast, the percentage of new domestic and imported light trucks with factory-installed rear window wipers has greatly increased since Ward's started reporting rear window wipers and are much larger than those for cars. Also note that the percentages of light trucks with rear window wipers are very similar to the percentages of light trucks with rear window defoggers. In general, almost all light trucks with rear window defoggers also have rear window wipers. For model year 2001 light trucks, almost all minivans and SUVs had rear window wipers, but no pickup trucks or full-sized vans had rear window wipers.

Table 6
Percentage Factory-Installed Rear Wipers By Model Year And Vehicle Group

Model Year	Domestic Cars	Imported Cars	Domestic Light Trucks	Imported Light Trucks
1982	3.2%			
1983	2.6%	33.3%		2.6%
1984	3.0%	31.4%		8.8%
1985	3.5%	25.2%		13.2%
1986	4.6%	27.8%		10.9%
1987	5.0%	34.8%	10.6%	8.8%
1988	4.2%	23.0%	16.3%	11.7%
1989	2.8%	17.2%	18.2%	26.5%
1990	4.8%	15.2%	24.4%	21.2%
1991	3.6%	14.1%	27.3%	40.6%
1992	3.8%	11.6%	32.1%	44.9%
1993	5.7%	13.3%	38.4%	55.2%
1994	4.4%	11.1%	35.3%	63.5%
1995	5.1%	9.7%	38.6%	70.2%
1996	2.8%	11.6%	44.6%	86.7%
1997	4.3%	10.3%	37.8%	91.9%
1998	2.8%	10.1%	49.4%	94.9%
1999	2.8%	11.6%	46.4%	94.7%
2000	5.4%	10.9%	47.6%	97.0%
2001	4.4%	13.1%	51.5%	98.0%

Rear window defoggers were already standard equipment on most make-models of minivans and SUVs when these models were introduced, so the sample of vehicles without rear window defoggers is too small for meaningful results. Conversely, defoggers are so rare on pickup trucks and full-sized vans that the sample of pickup trucks and full-sized vans with rear defoggers is also too small for meaningful results. Similarly, except for those make-models that were always equipped with them, rear-window wipers are so rare that the samples of vehicles with the wipers will be too small for meaningful results. Therefore, the analysis estimates the effectiveness of rear-window defoggers for cars only, where there are numerous make-models that experienced a substantial or at least moderate year-to-year increase in the proportion equipped with defoggers.

A rear-window defogger file was created with the necessary information needed from the Ward's Automotive Yearbooks. The file contains the percentage of rear-window defoggers for cars by make, model, and model year for model years 1973-2001.

STATE CRASH DATA

A large sample of crash-involved cars is needed to maximize the possibility of detecting any effect of rear-window defoggers. Defoggers may reduce crashes where the driver is backing up and hits something with the rear of the car or when the driver is changing lanes and the car gets hit in the rear. A clear window will obviously help a driver who is backing up, and can also help a driver see if it is safe to change lanes. The goal is to detect if there are proportionately fewer backing up and changing lane crashes involving cars with rear-window defoggers than cars without rear-window defoggers. Since about 50 percent of the model year 1980 domestic cars had rear-window defoggers, the data needed to include a large sample of cars without rear-window defoggers – i.e., 1975-1985 model year vehicles. The data also needed to include specific make, model and model year of vehicles so that the percentage of factory-installed rear window defogger information available in Ward's Automotive Yearbooks can be linked to the crash data. Therefore, the analyses require a large sample of crash data that includes vehicle maneuver (changing lanes or backing up vs. other), impact location (rear damage), and a large range of model years and VINs or other codes to identify the specific make, model, and model year of cars. State data files are the only data source available to NHTSA that can furnish an adequate number of cases for statistical analysis.

Although the State data files are the best source of data available for this analysis, the State data files may not include all backing up crashes. Backing-up crashes that cause little or no property damage may not be included in a State's crash file either because it was not reported to the police or it did not meet the State's minimum-reporting threshold. (As we shall see later on, Pennsylvania has very few backing-up crashes.)

The agency currently receives data from 17 States and maintains these data files for calendar years 1989 and onward for analysis. However, data from 13 States were not used in the analysis: North Carolina, because NHTSA does not have their files prior to 1992; California, Georgia, Illinois, Kansas, New Mexico, Ohio and Washington, because they do not specify the impact location (rear impact); Virginia, because it does not have vehicle make and model; Texas, because it does not have "changing lanes" as a possible vehicle maneuver; Missouri, Utah and Indiana, because earlier years of crash data were not easily available. Florida, Maryland, Michigan and Pennsylvania files were initially considered for the analysis. (As we shall see later on, only the Florida and Michigan data turned out to be suitable for the analysis.)

Previous calendar years of crash data for these four States were obtained to supplement the years currently available at NHTSA. Design Research Engineering supplied 1986–1988 Florida crash data files, the 1986-1988 Maryland crash data files, and the 1981-1988 Michigan crash data. The Pennsylvania Department of Transportation supplied the 1980-1988 Pennsylvania crash data files. Table 7 shows the calendar years of crash data used in the analysis by State. Michigan's 1992-2000 data files could not be used in the analysis because the make, model and model year were not available. Florida's 2000 data file was not used in the analysis because it does not contain low-severity non-injury crashes. Pennsylvania's 1980 file was not used because it would have contributed fewer than 50 cases to the main analysis.

Table 7
Calendar Years Of Crash Data By State

State	Calendar years
Florida	1986-1999
Maryland	1986-2000
Michigan	1981-1991
Pennsylvania	1981-2000

ANALYSIS DATABASES

The analysis databases are combinations of the rear-window defogger file and the State crash files. Linking the rear-window defogger file to the State crash files created the analysis files. The files were linked by matching the vehicle make, model, and model year for cars as defined in Ward's. In Florida, Maryland, and Michigan State data files, vehicle make and model was decoded from the Vehicle Identification Number (VIN); model year was not decoded from the VIN, but taken directly from the variable available on the State data files. In Pennsylvania for calendar years 1988-2000, make and model was decoded from the VIN; model year was not decoded from the VIN, but taken directly from the variable available on the State data file. For calendar years 1980-1987, very few VINs were included on the Pennsylvania data files, so make, model and model year were taken directly from the variables available on the State data files. The resulting analysis databases are vehicle-oriented files, with one record for each car that was involved in a crash. Since the rear-window defogger file only contains cars with 1973-2001 model years, the analysis files also only contain cars with 1973-2001 model years. The Michigan analysis file only has 1973-1992 model years and Florida only has 1973-2000 model years.

We believe rear window defoggers may have the potential to reduce crashes in which a driver was backing up or changing lanes before the crash. These tasks can be facilitated by vision through the rear window. Only cars with rear damage were included in the analysis. Consequently, rear impacts in which the vehicle maneuver was backing up or changing lanes were considered the only crashes that should be reduced as a result of rear window defoggers (i.e., treatment group). Vehicles with rear damage that were stopped prior to the crash were considered the control group. Since these vehicles were not moving at all and did nothing to cause the crashes, we can be relatively certain that rear-window defoggers would not have prevented them. Therefore, the critical parameters besides the presence or absence of rear window defoggers that must be defined in each State file are rear damage and vehicle maneuver (backing up and changing lanes vs. stopped).

Every State has its own unique ways of coding a vehicle's impact location. "Rear damage" cannot be defined exactly the same in each State, but at least the definitions can be made as similar as possible. Rear damage was defined to include any impact into the rear portion of the car where the driver might at least have had a chance to see the other vehicle behind it, either indirectly (via rear view mirror) or directly through the rear window. This includes rear-corner impacts and in some cases even impacts resulting in damage to the back portion of the side of the car, because sometimes when a car changes lanes into the path of another vehicle traveling in the same direction damage to the side of the car is possible. Given this extended definition of "rear damage," the following cars were included in the analysis:

State	Definition	Percent of Cars* that have Rear Damage
Florida	Impact = 6, 7, 8, 9, 10	26.6
Maryland 1986-1992	Impact = 7, 8, 9	26.7
Maryland 1993-2000	Impact = 7, 8, 9, 10, 11,12	25.7
Michigan	Impact = 4, 5, 6	24.4
Pennsylvania	Impact = 4, 5, 6, 7, 8	20.8

* 1973 – 2001 model year domestic cars with rear defogger information.

Every State also has its own unique ways of coding "backing up," "changing lanes," and "stopped" maneuvers. Florida, Maryland, and Pennsylvania have a variable called Veh_man1 that indicates the vehicle's maneuver just prior to impact. Michigan has a "driver intent" variable that indicates the driver's intended maneuver just prior to impact. The following shows the definition for "backing up," "changing lanes" and "stopped" by State:

State	Maneuver	Definition
Florida	Changing lanes	Veh_man1 = 6
Florida	Backing	Veh_man1 = 4
Florida	Slowing, stopped/stalled*	Veh_man1 = 2
Maryland	Changing Lanes	Veh_man1 = 7
Maryland	Backing	Veh_man1 = 11
Maryland	Stopped in Traffic Lane	Veh_man1 = 6
Michigan	Changing lanes	Drintent = 3
Michigan	Backing	Drintent = 11
Michigan	Stopped on Road	Drintent = 12
Pennsylvania	Changing Lanes to left and to right	Veh_man1 = 2,3
Pennsylvania	Backing Up	Veh_man1 = 1
Pennsylvania	Stopped	Veh_man1 = 24

*Includes some vehicles that were slowing rather than stopped. Both "slowing" and "stopped" vehicles may be considered part of the control group, since vision through the rear window is not an issue.

If rear window defoggers were effective, they would reduce crashes in which a driver is backing up and changing lanes – relative to the number that would have been expected if the cars had not been equipped with rear window defoggers. There is no direct way to count the crashes that were prevented, nor is there any way to determine if the rear window defogger was activated during the pre-crash maneuver. The basic approach was to study the change in proportion of crashes that were relevant, assuming that the presence or absence of defogger does not affect the occurrences of non-relevant crashes. The analytical methodology chosen for this study controls for some demographic characteristics of the drivers along with environmental and vehicle factors.

Ideally, only make-models that did not offer rear window defoggers as optional equipment in an earlier model year but offered them as standard equipment on the next model year should be analyzed. But there were no make-models where this happened. Another ideal group to analyze would have been make-models that had a big increase in the percentage of defoggers in consecutive model years. But there were very few make-models that had an increase of 50 percentage points or more and only about 30 make-models had an increase of 20 percentage points or more in consecutive model years.

In order to have a large sample of cars, the sample included make-models with a large volume of crashes: make-models that had at least 500 cases in Pennsylvania where the cars were changing lanes, backing up or stopped just prior to impact and had rear damage. Pennsylvania was used because there were 20 calendar years of data. This ensured the sample of vehicles would include a wide variety of model years that have various percentages of defoggers. If Florida or Maryland had been used then the sample would have been mostly more recent model years that have a high percentage of cars with defoggers. If Michigan had been used then the sample would have been mostly earlier model years that had very few make-models with defoggers.

But since only 1981-1991 CY of data were available in Michigan, the mix of cars with a large volume of crashes was different in Michigan than in Pennsylvania, so Michigan's database included 34 additional make-models not included in the other States' databases. These additional make-models had at least 750 crashes in Michigan. Appendix A contains a list of the make-models included in the analysis. The list includes for each make-model the percentage of rear window defoggers by model year and the number of crashes in each State. Appendix A also contains a list of additional make-models included in Michigan's database.

Table 8 shows the percentage of defoggers by model year for six make-models. Some make-models that met the above criteria were excluded from the analysis because the percentage of defoggers jumped all over the place over the span of several model years. The fluctuating percentages are unusual, since for most cars the percentage gradually increases by model year. Intuitively, it appears that some of the reported percentages could be erroneous, but without another source of percentages it is hard to know. So these make-models were excluded from the analysis. For example, the Chevrolet Camaro was excluded from the analysis for this reason (see Table 8). From model year 1982 to 1987, about 60 percent of Camaros had rear defoggers. Then the percentage fluctuated between model year 1988 and 1997. During that time, it dropped into the 50's; plummeted to a low of 23.4 percent in model year 1990; rose to the 56.7, 62.3, and 78.4; and dropped again into the 60's before it was consistently in the 70's.

Table 8
Percentage Of Rear Window Defoggers By Model Year For Some Make-Models

Model Year	Ford Escort (1981-90 design)	Chevrolet Caprice (1977-96 design	Ford Crown Victoria (1979 - design)	Ford Taurus (1986-95 design)	Chevrolet Citation	Chevrolet Camaro (1982 - design)
1977		32.0%				
1978		34.5%				
1979		40.6%	38.5%		43.7%	
1980		45.8%	39.1%		45.8%	
1981	36.5%	45.9%	61.2%		45.0%	
1982	38.3%	50.1%	64.5%		47.2%	61.8%
1983	35.9%	57.0%	75.3%		53.4%	63.6%
1984	35.8%	60.7%	63.3%		63.6%	65.8%
1985	42.2%	63.5%	80.1%		61.1%	66.3%
1986	60.0%	70.5%	80.9%	91.1%		68.9%
1987	67.5%	76.7%	86.7%	89.2%		62.5%
1988	80.6%	72.6%	89.3%	93.2%		55.0%
1989	86.5%	77.1%	87.5%	89.7%		53.8%
1990	88.0%	39.4%	89.4%	91.5%		23.4%
1991		81.9%	94.0%	100%		56.7%
1992		87.6%	95.3%	99.6%		62.3%
1993		88.8%	97.1%	100%		78.4%
1994		91.7%	97.7%	100%		69.7%
1995		88.3%	100%	100%		63.6%
1996		98.2%	100%			65.4%
1997			100%			64.0%
1998			100%			78.8%
1999			100%			73.9%
2000			100%			73.2%
2001			100%			89.1%
Increase	51.5	66.2	61.5	8.9	17.4	27.3

Model years where the percentages were much higher or lower than expected relative to previous and subsequent percentages were excluded from the analysis. These data may be correct or erroneous, but since there is no way to know, they were excluded from the analysis. Model year 1990 for the Chevrolet Caprice was excluded from the analysis for this reason (see Table 8). The percentage of defoggers steadily increased from 32.0 percent in 1978 to 98.2 percent in 1996 except for a severe drop to 39.4 percent in 1990.

Table 9 shows the number of cases in the analysis databases by crash type and State. At this point, the analyses were limited to Florida and Michigan because they have larger samples of total cases and far more changing lane and backing cases. Maryland has fewer than 75,000 total cases. Although Pennsylvania has almost 175,000 total cases, only 4 percent are changing lane and backing cases.

Table 9
Number Of Cases in Analysis Databases By Crash Type And State

State	Changing Lane & Backing Crashes	Stopped Crashes	Total
Michigan	71,668	250,158	321,826
Florida	44,761	276,357	321,118
Pennsylvania	7,791	174,669	182,460
Maryland	13,121	61,604	74,725

ANALYSIS METHOD

To detect if rear window defoggers are effective, we considered and tested several different hypotheses. The first is that rear window defoggers reduce relevant crashes all the time. In other words, just the presence of rear window defoggers reduces backing up and changing lanes crashes. This hypothesis is probably unrealistic since rear window defoggers only improve rearward vision under certain conditions: when drivers turn them on because there is condensation, frost, ice, and/or snow on the back window. Nevertheless, this hypothesis was tested. (In part, this hypothesis had to be included in each model for methodological reasons, as will be explained later on.)

The second hypothesis is that rear window defoggers reduce relevant crashes when they are used. There is no way to determine if the rear window defogger was in use prior to the crash on the analysis databases, but we surmise that they are most likely used if any one or more of the following adverse conditions occurred: snowing or raining, during the earlier part of the morning, or during winter.

A third hypothesis is that rear window defoggers have independent (and additive) effects of reducing relevant crashes under the three individual conditions when they are most likely used: raining or snowing, during early morning, or during the winter. This hypothesis suggests the overall effect of rear window defoggers is the sum of the effects when it is raining or snowing, during early morning, and during the winter.

The fourth hypothesis is that the more adverse the conditions, the more effective the defoggers are at reducing relevant crashes but that the effect of each type of adverse condition is equal. In other words, when it is raining, morning, and winter, the effect of defoggers is three times as large as when just one of these conditions is present. This hypothesis is very similar to the third hypothesis in that is assumes that the effect is additive, but it differs in that it assumes each type of adverse condition has the same effect.

This additive effect assumed by the third and fourth hypotheses is unlikely since there is probably not an additional benefit when there is more than one adverse condition. We believe rear window defoggers should have a benefit when they are being used and they will be used

14

if any one adverse condition is present. For example, if it is raining then they are most likely being used. We do not believe that they are even more likely to be used if it is raining and it is also during the early morning. The third and fourth hypotheses were also tested (for the sake of completeness) even though we believe the second hypothesis is the most reasonable.

Logistic regression was used to estimate the effect of defoggers on the probability that the crash was relevant (as opposed to being a control group crash), while controlling for other factors. Estimating the impact of defoggers in reducing relevant crashes could be confounded by factors related to the driver, environment, vehicle, crash or other circumstances. To accurately estimate the impact of rear window defoggers, variables were included in the logistic regression to control for those factors, other than defoggers, which could influence the proportion of relevant crashes. For example, if rear window defoggers are in newer cars, or are more likely to be driven by older drivers than by other segments of the driving population, then driver and vehicle characteristics could confound estimating the impact of rear window defoggers. As a result, the age and sex of the driver, whether or not the crash occurred during adverse weather conditions (while raining or snowing when rear window defoggers are more likely to be used), whether or not the crash occurred during the morning or winter (also when rear window defoggers are more likely to be used), the age of the vehicle, vehicle make-model, and calendar year of the data were chosen for inclusion in the logistic regression model.

The following regression model was run on the Florida data set to test our second hypothesis – defoggers reduce relevant crashes in conditions when they are most likely to be used:

MODEL CLBKSTOP = DEF USED DEF USED ADWEA WINTER MORN VEHAGE
VEHAGE2 CY86 CY87 CY88 CY89 CY90 CY91 CY92 CY93
CY94 CY95 CY96 CY97 CY98 DRVMALE M14_30 M30_50
M50_70 M70+ F14_30 F30_50 F50_70 F70+ CAVALIER1
ESCORT1...TERCEL;

CLBKSTOP is a flag that indicates whether the crash-involved car was changing lanes or backing (failures) or stopped (successes). All records where the crash-involved car was changing lanes or backing have CLBKSTOP = 1. All records where the crash-involved car was stopped have CLBKSTOP = 2.[2]

DEF is the percentage of rear window defoggers for that make-model in that model year. It is a continuous variable with values from 0 to 100.

DEF_USED is an interaction variable that expresses the probability that the rear window defoggers were most likely being used prior to the crash. It has the value of DEF if it is snowing or raining, during the earlier part of the morning or during winter. DEF_USED = DEF * USED (defined below). For example, if 1989 Ford Taurus (89.7 percent had rear window defoggers) had a collision when it was raining, then DEF = 89.7 and DEF_USED = 89.7. DEF_USED has the value of zero if it is unlikely that the rear window defogger was in use. DEF_USED = 0 when it is not snowing, not raining, not during the earlier part of the

---

[2] SAS/STAT® User's Guide, Version 6, Fourth Edition, Volume 2, SAS Institute, Cary, NC, 1989, pp.1071-1126. The LOGISTIC procedure in SAS prefers values of 1 for failures and 2 for successes.

morning and not during winter even if the car is equipped with rear window defoggers. For example, if a 1989 Ford Taurus had a crash that occurred on a bright, sunny afternoon in May, then DEF_USED = 0. DEF_USED also equals zero when the car does not have rear window defogger even if it is snowing or raining, during the early morning or during winter.

USED indicate if the conditions when rear window defoggers are most likely being used. It has the value one if it is snowing or raining, during the earlier part of the morning or during winter. It equals zero, otherwise. Specifically, USED = 1 if ADWEA = 1 or MORN = 1 or WINTER = 1 (defined below).

ADWEA has the value 1 if the crash occurred when it was raining or foggy, 0 otherwise.

WINTER has the value 1 if the crash occurred during January thru April or November or December, 0 if it occurred in May-October.

MORN has the value 1 if the crash occurred during 6:00 am to 9:59 am.

The model included both a linear and non-linear variable to account for vehicle age. The linear variable (VEHAGE) is age of the vehicle when the crash occurred (CY - MY). The non-linear variable (VEHAGE2) is VEHAGE * VEHAGE.

All the "CY" variables are indicator variables for calendar year, they have the value 1 if true otherwise the value of 0. For example, CY86 has value 1 if the calendar year is 1986, 0 otherwise.

The model includes one dichotomous and eight continuous variables to express driver age and gender. Kahane in the *Vehicle Weight, Fatality Risk and Crash Compatibility of Model Year 1991-99 Passenger Cars and Light Trucks*[3] used this approach. The dichotomous variable is DRVMALE. It has the value 1 if the driver is male and 0 if the driver is female.

> "Driver age is expressed as a 4-piece linear variable, separately for males and females (eight variables in all): four connected straight-line segments, one from age 14 to 30, another from 30 to 50, another from 50 to 70, and the last from 70 and up. The eight variables are:
>
> M14_30 = 30 – DRVAGE for male drivers age 14-30, otherwise it is 0 for male drivers age 30+ and all female drivers.
>
> M30_50 = 50-DRVAGE for male drivers age 30-50; = 20 for male drivers age 14-30; or =0 for male drivers age 50+ and for all female drivers.
>
> M50_70 = DRVAGE –50 for male drivers age 50-70; = 20 for male drivers70 +; =0 for male drivers age 14-50 and all female drivers.

---

[3] Kahane, C.J., *Vehicle Weight, Fatality Risk and Crash Compatibility of Model Year 1991-99 Passenger Cars and Light Trucks*, NHTSA Technical Report No. DOT HS 809 662, Washington, 2003, pp. 67-74.

M70+ = DRVAGE −70 for male drivers age 70+; =0 for male drivers age 14-70 and all female drivers.

F14_30 = 30 − DRVAGE for female drivers age 14-30, otherwise it is 0 for female drivers age 30+ and all male drivers.

F30_50 = 50-DRVAGE for female drivers age 30-50; = 20 for female drivers age 14-30; or =0 for female drivers age 50+ and for all male drivers.

F50_70 = DRVAGE −50 for female drivers age 50-70; = 20 for female drivers70 +; =0 for female drivers age 14-50 and all male drivers.

F70+ = DRVAGE −70 for female drivers age 70+; =0 for female drivers age 14-70 and all male drivers.

For example, a 40-year-old male driver would have M30-50 = 10, and the other variables set to zero. A 25-year-old male driver would have M30_50 = 20, M14_30 = 5, and the others set to zero. Conversely, a 60-year-old female driver would have F50_70 = 10 and the others set to zero. A 75-year-old female driver would have F50_70 = 20, F70+ = 5, and the others set to zero.

The rationale for defining the variables that way is that it treats 50 years as the baseline age. Each year that a driver is younger than 50 has some effect (usually increasing) on fatality risk, and each year that a driver is older that 50 has another effect (also usually increasing)."[4] Given any specific age and gender, there is exactly one combination of the nine variables that will indicate a driver of that age and gender. These variables allow different linear relationships between age and crash risk in different age/gender groups.

Indicator variables for each make model were included in the model. These indicator variables are needed because some make-models are driven differently than others. For specific make-models, there may be two or more indicator variables included in the model depending on whether or not the car had a major redesign. For example, the Chevrolet Cavalier has two indicator variables: CAVALIER1 and CAVALIER2. The Chevrolet Cavalier was redesigned in model year 1995. So CAVALIER1 has the value 1 if the car was a 1983-1994 model year Chevrolet Cavalier, 0 otherwise and CAVALIER2 has the value 1 if the car was a 1995-2000 model year Cavalier, 0 otherwise. Similar indicator variables were made for all the other cars included in the model. The reference car was the 1975-1978 Volkswagen Rabbit.

---

[4] Kahane, C.J., *Vehicle Weight, Fatality Risk and Crash Compatibility of Model Year 1991-99 Passenger Cars and Light Trucks*, NHTSA Technical Report No. DOT HS 809 662, Washington, 2003, pp. 69-70.

HYPOTHESES –
- **REAR WINDOW DEFOGGERS REDUCE RELEVANT CRASHES IN THE PRESENCE OF ANY ONE OR MORE CONDITIONS WHEN THEY ARE MOST LIKELY USED (but the effect does not increase if there is more than one adverse condition) and**
- **REAR WINDOW DEFOGGERS REDUCE RELEVANT CRASHES ALL THE TIME**

This model contained two independent variables that indicate if rear window defoggers are effective. DEF_USED indicates if the presence of rear window defoggers when they are most likely to be used affects the rate of changing lane and backing crashes. DEF indicates if the mere presence of rear window defoggers affects the rate of changing lane and backing crashes. In other words, DEF will indicate if rear window defoggers are effective all of the time and DEF_USED will indicate if rear window defoggers are effective in the conditions when they are most likely to be used. Thus, the DEF_USED coefficient in the model will test our hypothesis that defoggers reduce relevant crashes when they are most likely used. But the DEF coefficient must also be included in the logistic regression (because interaction terms such as DEF_USED should not be included without also including the main effects DEF and USED). DEF will indicate if defoggers when present affect relevant crashes. We believe that just the presence of rear window defoggers should not have an effect on relevant crashes. If our hypothesis is correct that defoggers reduce relevant crashes when used and they have no effect when they are not used, then the DEF_USED coefficient should be negative and significant and the DEF coefficient should be close to zero. In fact, even if the DEF coefficient were substantial, we suspect it more likely indicates a possible bias in the model or data than a real effect of defoggers. We believe DEF_USED is the more important coefficient.

Table 10 shows the results of the logistic regression for the 1986-1999 Florida data. The coefficient for DEF_USED is –0.00002, in the "right" direction but not statistically significant (Chi-Square = 0.002).

Technically, the DEF_USED coefficient represents the change in the log odds ratio of relevant (changing lanes and backing) to non-relevant (stopped) crashes when defoggers are most likely used for a 1 percent increase in the percentage of cars with rear window defoggers. A negative coefficient represents a reduction that is associated with the presence of rear window defoggers when most likely used. Thus, a 100 percent increase in the percentage of cars with rear window defoggers is associated with 0.002 reduction in the log of relevant crash rate when defoggers are most likely used. The coefficient can be translated into the percentage change in the expected number of relevant crashes in the following way:

Expected percentage effectiveness = 100*[1-exp (DEF coefficient*100)].

In other words, cars with rear window defoggers when they are most likely used have 100 * [1- (exp (.002))] = 0.2 percent reduction in changing lanes and backing crashes relative to crashes where the vehicle was stopped. Rear window defoggers when most likely used have no effect on relevant crash, since the chi-square value is 0.002 and chi-square needs to be at least 3.89 for statistical significance at the 0.05 level.

The coefficient for DEF is 0.00162, in the "wrong" direction but also non-significant (Chi-Square = 3.184). Thus, the Florida data do not show any statistically significant effect for defoggers.

Table 10
1986-1999 Florida
Cars With Rear Impacts That Were Changing Lanes, Backing Up, Slowing,
Stopped, Or Stalled

Parameter	Estimate	Standard Error	Wald Chi-Square	Pr > ChiSq
Intercept	-1.6579	0.1343	152.2927	<.0001
DEF	0.00162	0.000909	3.184	0.0744
DEF_USED	-0.00002	0.000435	0.002	0.9647
USED	-0.1511	0.0416	13.1992	0.0003
ADWEA	-0.7571	0.0234	1051.1949	<.0001
WINTER	0.0518	0.0233	4.9489	0.0261
MORN	-0.0985	0.0196	25.165	<.0001
VEHAGE	0.0173	0.00427	16.3514	<.0001
VEHAGE2	0.000608	0.000225	7.3093	0.0069
CY86	0.0273	0.0448	0.3715	0.5422
CY87	-0.1325	0.0426	9.6577	0.0019
CY88	-0.1503	0.0399	14.2253	0.0002
CY89	-0.1073	0.0373	8.2893	0.004
CY90	-0.1011	0.0361	7.83	0.0051
CY91	-0.0766	0.0347	4.8868	0.0271
CY92	-0.0459	0.0327	1.9701	0.1604
CY93	-0.0979	0.0319	9.4068	0.0022
CY94	-0.1121	0.0306	13.4317	0.0002
CY95	0.013	0.028	0.2165	0.6417
CY96	-0.2027	0.0284	50.9083	<.0001
CY97	-0.0186	0.0264	0.4957	0.4814
CY98	0.0373	0.0254	2.1556	0.142
DRVMALE	-0.0285	0.0296	0.9264	0.3358
M14_30	0.0502	0.0022	521.6789	<.0001
M30_50	0.0175	0.00151	134.1712	<.0001
M50_70	0.0276	0.00168	269.659	<.0001
M70_96	0.0996	0.00282	1243.4684	<.0001
F14_30	0.0549	0.0024	520.9121	<.0001
F30_50	0.000669	0.0016	0.1753	0.6755
F50_70	0.0471	0.00182	671.5062	<.0001
F70_96	0.0886	0.00359	609.8991	<.0001
CAVALIER1	-0.8086	0.1146	49.7464	<.0001
ESCORT1	-0.8556	0.1179	52.6484	<.0001
CAPRICE1	0.0319	0.1123	0.0805	0.7766
TAURUS1	-0.6526	0.1212	28.979	<.0001
TEMPO	-0.6629	0.1172	31.9908	<.0001

19

Table 10 – Continued

Parameter	Estimate	Standard Error	Wald Chi-Square	Pr > ChiSq
CELEBRITY	-0.7663	0.1196	41.0775	<.0001
GRANDAM	-0.8848	0.1185	55.708	<.0001
CIERA	-0.8502	0.12	50.1573	<.0001
CUTLASS1	-0.5913	0.1166	25.7121	<.0001
CHEVETTE	-0.4266	0.1243	11.7865	0.0006
CENTURY1	-0.845	0.1217	48.2322	<.0001
MUSTANG	-0.6024	0.1145	27.6735	<.0001
RELIANT	-0.8532	0.1263	45.6675	<.0001
CORSICA	-0.6775	0.1247	29.5101	<.0001
SUNBIRD	-0.8745	0.1272	47.2302	<.0001
ESCORT2	-0.8156	0.1234	43.6733	<.0001
MONTE1	-0.5853	0.1183	24.4601	<.0001
CROWNVIC	0.3289	0.1175	7.8379	0.0051
LUMINA	-0.6828	0.1272	28.8221	<.0001
CITATION	-0.5525	0.141	15.3483	<.0001
ARIES	-0.8317	0.1248	44.4337	<.0001
FAIRMONT	-0.4133	0.12	11.8572	0.0006
MALIBU	-0.5068	0.124	16.6945	<.0001
ACCORD1	-0.8565	0.1264	45.8754	<.0001
REGAL	-0.5897	0.1188	24.6408	<.0001
P6000	-0.638	0.1287	24.574	<.0001
DELTA1	-0.1761	0.1193	2.1787	0.1399
GMARQUIS	-0.5362	0.1215	19.4728	<.0001
OMNI4DR	-0.7307	0.163	20.0868	<.0001
SABLE	-0.7579	0.1309	33.5421	<.0001
LESABRE1	-0.9783	0.1276	58.7781	<.0001
TOPAZ	-0.6658	0.1337	24.7831	<.0001
SHADOW	-0.8883	0.1365	42.3473	<.0001
SUNDANCE	-0.8107	0.1395	33.7796	<.0001
DEVILLE1	-0.3088	0.1219	6.4152	0.0113
DEVILLE2	-0.4892	0.1261	15.0633	0.0001
BERETTA	-0.5711	0.1292	19.5558	<.0001
THUNDER1	-0.6749	0.1272	28.1715	<.0001
COUGAR1	-0.6792	0.1297	27.4176	<.0001
SKYLARK1	-0.592	0.138	18.399	<.0001
DEVILLE3	-0.6975	0.1208	33.3294	<.0001
HORIZON	-0.6784	0.1704	15.8464	<.0001
SATURN	-1.1415	0.1323	74.4713	<.0001
CAVALIER2	-1.107	0.1426	60.27	<.0001
SKYLARK2	-0.943	0.1324	50.7292	<.0001
CALAIS	-0.9328	0.1331	49.0969	<.0001
DELTA2	-0.8705	0.1318	43.6156	<.0001
GRANDPRIX1	-0.8488	0.1348	39.6177	<.0001

Table 10 – Continued

Parameter	Estimate	Standard Error	Wald Chi-Square	Pr > ChiSq
CIVIC1	-0.9198	0.1254	53.8029	<.0001
LESABRE2	-0.3507	0.1229	8.1429	0.0043
GRANADA	-0.1958	0.1293	2.2906	0.1302
TOWNCAR	-0.4348	0.1223	12.6429	0.0004
ACCORD2	-1.0276	0.1292	63.2644	<.0001
ACCORD3	-0.791	0.1275	38.4732	<.0001
SENTRA1	-1.1258	0.127	78.6326	<.0001
CAMRY1	-0.8883	0.1268	49.0939	<.0001
LTD	0.0808	0.1248	0.4191	0.5174
VOLARE	-0.2067	0.129	2.5661	0.1092
CAPRICE2	-0.0625	0.1284	0.2372	0.6262
CIVIC2	-1.1626	0.1349	74.2177	<.0001
ACCLAIM	-0.7761	0.1489	27.1552	<.0001
GRANDPRIX2	-0.339	0.1216	7.7791	0.0053
CUTLASS2	-0.0547	0.1281	0.1827	0.6691
SUPREME1	-0.6846	0.1463	21.8879	<.0001
SKYHAWK	-0.7723	0.156	24.5135	<.0001
ELECTRA	-1.14	0.1385	67.7382	<.0001
HORNET	-0.0827	0.1386	0.3563	0.5506
MONTE2	-0.1587	0.136	1.3605	0.2435
BONNEVILLE	-0.1353	0.1315	1.0581	0.3037
PARISIENNE	-0.3587	0.1503	5.6926	0.017
NOVA1	-0.2251	0.1257	3.2076	0.0733
CAMARO	-0.4467	0.1218	13.4537	0.0002
PINTO	-0.019	0.1439	0.0175	0.8948
LEBARON1	-0.8419	0.1371	37.7211	<.0001
CAMRY2	-0.6545	0.1312	24.881	<.0001
ASPEN	-0.3034	0.129	5.5308	0.0187
TAURUS2	-0.7342	0.1473	24.8572	<.0001
ALTIMA	-0.921	0.1314	49.1283	<.0001
DAYTONA	-0.8438	0.1475	32.7431	<.0001
THUNDER2	-0.3195	0.1313	5.9238	0.0149
CHEVELLE	0.1859	0.1463	1.6149	0.2038
SPIRIT	-0.7976	0.1496	28.432	<.0001
LEGACY	-1.1856	0.2229	28.2828	<.0001
PROBE	-0.8096	0.1453	31.0575	<.0001
OLDS98	-0.9531	0.1377	47.9346	<.0001
LEBARON2	-0.6259	0.139	20.2716	<.0001
DYNASTY	-0.819	0.1529	28.7071	<.0001
MONZA	-0.4225	0.1687	6.2762	0.0122
EAGLE	-0.1943	0.338	0.3304	0.5654
OMEGA	-0.4624	0.1804	6.5732	0.0104
NEONPLY	-0.7177	0.1925	13.9012	0.0002

Table 10 – Continued

Parameter	Estimate	Standard Error	Wald Chi-Square	Pr > ChiSq
NEONDOD	-0.8636	0.1769	23.8328	<.0001
ZEPHYR	-0.4222	0.1492	8.0099	0.0047
FIREBIRD	-0.4837	0.1298	13.8907	0.0002
CENTURY2	-0.3467	0.1546	5.0307	0.0249
THUNDER3	-0.855	0.1361	39.4743	<.0001
INTREPID	-0.5406	0.1558	12.0399	0.0005
MARQUIS	-0.5115	0.1399	13.37	0.0003
COUGAR2	-0.753	0.1433	27.5981	<.0001
LEBARON3	-0.2955	0.1508	3.8413	0.05
CORDOBA	-0.2765	0.1558	3.1491	0.076
NEWYORKER	-0.899	0.1519	35.033	<.0001
DART	-0.1866	0.1528	1.4905	0.2221
DIPLOMAT	0.3079	0.1245	6.111	0.0134
CONTOUR	-0.8677	0.1639	28.0303	<.0001
FIRENZA	-0.7187	0.1702	17.8207	<.0001
SUPREME2	-0.5465	0.1229	19.7705	<.0001
LEGACY1	-0.6543	0.3883	2.839	0.092
NOVA2	-0.8272	0.1522	29.5268	<.0001
COUGAR3	-0.1681	0.143	1.382	0.2398
VALIANT	-0.2049	0.2075	0.9752	0.3234
GREMLIN	-0.3648	0.1803	4.094	0.043
JETTA	-1.0479	0.1388	56.9669	<.0001
COROLLA	-1.0354	0.1256	67.916	<.0001
LOYALE	-0.9161	0.1702	28.9677	<.0001
ACCORD4	-0.9514	0.1339	50.4566	<.0001
VOL240	-1.068	0.1382	59.7267	<.0001
EXCEL	-0.9987	0.1326	56.6941	<.0001
MAXIMA1	-0.8257	0.1376	36.0173	<.0001
SUBARU	-0.8949	0.1716	27.2023	<.0001
SPECTRUM	-1.0584	0.1526	48.0771	<.0001
GOLF	-1.0393	0.161	41.6642	<.0001
TERCEL	-1.0464	0.195	28.7966	<.0001
SENTRA2	-0.9259	0.1374	45.4432	<.0001
RABBIT	-1.1188	0.1591	49.4222	<.0001
MAXIMA2	-0.8316	0.1391	35.7504	<.0001
CELICA	-0.6559	0.1372	22.8457	<.0001

The regression coefficient (0.0173) for VEHAGE shows that changing lanes and backing crashes increase relative to stopped involvements, as cars get older. Changing lanes and backing crashes increase 2 percent for every year a car gets older. The negative regression coefficient for almost all of the CY terms implies that changing lanes and backing crashes were less common in the past than in recent years. VEHAGE, VEHAGE2, and most of the CY terms are included in the model because they are significant.

The coefficients of the other independent variables seem reasonable. The positive coefficients for M14_30, M70+, F14_30, and F70+ show that the youngest and oldest drivers are especially prone to backing up or changing lane crashes. ADWEA and MORN are negative, indicating that changing lanes and backing crashes decrease relative to stopped involvements during adverse weather and early morning. USED is also negative, indicating the relevant crashes decrease during conditions when rear window defoggers are most likely used.

The make-model indicator variables are used only as control variables in the model. Some will have high or low coefficient values by chance alone indicating more or fewer changing lane and backing crashes than the average. We could reduce the number of these variables, if we grouped several make-models together. But there is no basis to group them. For example, not all small cars are driven in such a manner that they have fewer (or more) relevant crashes than large cars. Thus, all the individual make-model terms are included in the model and it is irrelevant if certain make-model terms are significant and others are not.

A similar model was run on the Michigan database. The adverse weather indicator variable for Michigan also had a value of 1 if the crash occurred when it was snowing, sleeting, hailing or freezing rain. Michigan's model included 34 additional make-models not included in the Florida model that had at least 750 crashes in Michigan.

The only other difference in the model was the number of individual CY terms. Since 1981-1991 Michigan data were analyzed, the model had 10 individual CY terms: CY81, CY82, …, CY90.

Table 11 shows the DEF_USED and DEF coefficients, percent reduction, and significance for the model by State. All of the results are non-significant indicating that rear window defoggers have no effect on changing lane and backing crashes, in all conditions and in conditions when they are most likely used. The DEF_USED coefficient in Michigan is in the "wrong" direction, corresponding to a 5 percent increase in changing lane and backing crashes, but this effect is not statistically significant.

Table 11
DEF_USED And DEF Coefficients And Percent Reduction By State

| State | DEF_USED | | DEF | |
	Coeff	Percent Reduction	Coeff	Percent Reduction
Michigan	0.00047	-5%	0.00015	-2%
Florida	-0.00002	0.2%	0.00162	-18%

HYPOTHESES –

- **REAR WINDOW DEFOGGERS REDUCE RELEVANT CRASHES AND THE EFFECTS ARE INDEPENDENT AND ADDITIVE FOR THE THREE ADVERSE CONDITIONS and**
- **REAR WINDOW DEFOGGERS REDUCE RELEVANT CRASHES ALL THE TIME**

The analysis was repeated to see if the individual adverse conditions when defoggers are most likely used reduce relevant crashes. This analysis tests three hypotheses:

1. rear window defoggers reduce relevant crashes when raining or snowing,

2. rear window defoggers reduce relevant crashes during the early morning, and

3. rear window defoggers reduce relevant crashes during the winter.

The model included three individual interaction variables, DEFADW, DEFMORN, and DEFWIN, instead of DEF_USED and USED.

DEFADW has the value of DEF if it is snowing or raining. DEFADW = DEF * ADWEA. For example, if 1989 Ford Taurus (89.7 percent had rear window defoggers) had a collision when it was raining, then DEF = 89.7 and DEFADW = 89.7. DEFADW has the value of zero if it was not raining or not snowing. It also has the value of zero if the make-model was not equipped with any defoggers in that model year.

DEFMORN has the value of DEF if it is during the early morning, particularly between 6:00 am and 9:59 am. DEFMORN = DEF * MORN. DEFMORN has the value of zero when the crash occurred before 6:00 am or after 10:00am.

DEFWIN has the value of DEF if it is during winter: November through April. DEFWIN = DEF * WINTER. DEFWIN has the value of zero when the crash occurred in the months May through October.

DEFADW will indicate if rear window defoggers reduce relevant crashes when it is raining or snowing. DEFMORN will indicate if rear window defoggers reduce relevant crashes during 6:00 am and 10:00 am. DEFWIN will indicate if rear window defoggers reduce relevant crashes during the winter.

Similar to the previous model, DEF must be included in the model because logistic regression requires the main effect to be included in the model when interaction terms are included. If our hypothesis is correct that rear window defoggers reduce relevant crashes under these three conditions, then DEFADW, DEFMORN, and DEFWIN coefficients should be negative. The regression model considers the effects to be linear and additive. If the effects are intrinsically not linear and not additive (e.g., if our second hypothesis is correct that the full benefit of defogger is achieved from any one adverse condition), then the model may spuriously calibrate some coefficients positive and others negative.

Table 12 shows the results and in fact, some of the coefficients are positive and some are negative. When raining or snowing, the coefficients are negative but not statistically significant. During the early morning, the coefficients are positive indicating an increase in relevant crashes. Michigan's coefficient is statistically significant, but Florida's is not. The coefficients for rear window defoggers during the winter are mixed. Michigan's coefficient is negative and Florida's is positive, but neither is significant. In any case, none of the coefficients is significant in the "right" direction, and we cannot conclude that rear window defoggers reduce relevant crashes when it is raining or snowing, during the early morning, or during the winter.

Table 12
DEFADW, DEFMORN, DEFWIN And DEF Coefficients And
Percent Reduction By State

State	DEFADW		DEFMORN		DEFWIN		DEF	
	Coeff	% Reduced	Coeff	% Reduced	Coeff	% Reduced	Coeff	% Reduced
MI	-0.00044	4%	0.00153	-17%*	0.000323	-3%	0.000122	-1%
FL	-0.00094	9%	0.000668	-7%	-0.00006	1%	-0.00061	6%

* Significant

HYPOTHESES –

- **REAR WINDOW DEFOGGERS REDUCE RELEVANT CRASHES AND THE EFFECTS ARE EQUAL AND ADDITIVE FOR THE THREE ADVERSE CONDITIONS and**
- **REAR WINDOW DEFOGGERS REDUCE RELEVANT CRASHES ALL THE TIME**

The analysis was repeated to test that the more adverse the conditions, the more effective are the defoggers at reducing relevant crashes. This may be because the defoggers are more likely to be used or that the defoggers improve the vision out of the rear window even more when there are two or more adverse conditions than when there is only one. We believe that if vision through the rear window is obscured, then the defoggers will be used whether one or more adverse conditions are present. It is probably unlikely that rearward vision can be improved more when more than one adverse condition is present. Either the driver will be able to see through the rear window or not. Nevertheless, this analysis tested to see if defoggers are effective when more adverse conditions are present.

The model included only one interaction variable, similar to the model with DEF_USED. The only difference was that this model used an interaction variable that escalated the effect of defoggers when more than one adverse condition existed. DEFAD is the weighted variable. It expresses the interaction between percent of rear window defoggers (DEF) and adverse conditions: adverse weather (ADWEA), winter (WINTER), and early morning (MORN). DEFAD = DEF * (ADWEA + WINTER + MORN). DEFAD is large when (according to this questionable hypothesis) there is a large probability that the rear window defoggers are being used. For example, if a vehicle equipped with a rear window defogger

25

had a collision that occurred during January at 8:00 AM when it was snowing, then ADWEA =1, WINTER = 1, MORN = 1, DEF = 100 and DEFAD = 100 * (1 + 1 +1) = 300. DEFAD is less when the probability is reduced. For instance, if a vehicle with a rear window defogger had a collision that occurred in May during the afternoon when it is raining, then DEFAD = 100. DEFAD is zero when there is little or no chance that rear window defoggers are being used. For example, if a vehicle with a rear window defogger had a crash that occurred on a bright, sunny afternoon in April, or a vehicle not equipped with a rear window defogger had a crash anytime, then DEFAD = 0.

The model included ADWEA, WINTER, and MORN, but it did not include a variable similar to USED. A similar variable to USED would have been the sum of ADWEA, WINTER, and MORN, but is not needed in the model since it is a linear combination of ADWEA, WINTER, and MORN.

Table 13 shows the results by State. All of the results in Table 13 are non-significant. Again, we conclude that these data fail to support any conclusion that rear window defoggers had an effect in either direction on changing lane and backing crashes.

Table 13
DEFAD And DEF Coefficients And
Percent Reduction By State

State	DEFAD		DEF	
	Coeff	Percent Reduction	Coeff	Percent Reduction
Michigan	0.000364	-4%	0.000121	-1%
Florida	-0.00001	0%	0.00162	-18%

CONCLUSIONS

The data do not show a benefit for rear window defoggers. The main analysis found that rear window defoggers have little or no effect on changing lane and backing crashes in conditions when they are most likely used (when it is raining or snowing, during the earlier part of the morning, or during winter).

Our inability to know whether a defogger was in use complicated but did not preclude our ability to measure their effectiveness. Our analytical approach accounts for this inability by comparing relevant to non-relevant crashes that are not affected by the use of rear window defoggers. For example, NHTSA's evaluations of antilock brake systems and air bags used a similar approach (i.e., the data did not indicate if ABS was activated or an air bag deployed) and found statistically significant effects.

It may be that rear windows defoggers, although convenient and helpful for drivers, are not that essential for safety. Perhaps, when drivers do not have the defoggers, they somehow compensate for their absence by relying more on their outside mirrors, using their turn signals more, making their lane changes more gradual, or simply not backing up or changing lanes when they cannot see what is behind them. Whatever the reason, we have not shown that rear window defoggers significantly reduced changing lane and backing crashes. Nevertheless, we would still expect most drivers to like rear window defoggers because they improve rearward vision. They can reduce or eliminate the need to scrape the rear window from outside the vehicle after a frost, a snowfall, or freezing rain, and these systems can eliminate the need for periodic stopping while driving to scrape accumulating snow or wipe condensation from the rear window. Even if drivers can successfully compensate for the lack of defoggers by avoiding lane changes and backup moves unless they are absolutely necessary, they would undoubtedly feel more confident, mobile, and secure if they had defoggers to give them a better view of the traffic behind them.

PRTCH is the difference between percentage of rear defoggers for the last and first model year for that make-model.

The State Columns show the number of rear impacts that were stopped, changing lanes, and backing in that State.

Shaded cells were excluded from the analysis because reported percentages were much higher or lower than expected relative to previous or subsequent percentages.

CG and MM2 are codes used in NHTSA evaluations for indicating the basic car group and specific make-model.

www.ingramcontent.com/pod-product-compliance
Lightning Source LLC
Chambersburg PA
CBHW081409170526
45166CB00010B/3275